The Wonderful
Wonder of Wonders

by

Jonathan Swift

Edited by Alessandro Gallenzi

ONEWORLD
CLASSICS

ONEWORLD CLASSICS LTD
London House
243-253 Lower Mortlake Road
Richmond
Surrey TW9 2LL
United Kingdom
www.oneworldclassics.com

First published by Oneworld Classics Limited in 2009

Printed in Great Britain by TJ International, Padstow, Corwall

ISBN: 978-1-84749-132-9

Contents

The Wonderful
Wonder of Wonders

The Grand Mystery

or

Art

of Meditating over a
House of Office
Restored and Unveiled
After the Manner of

The Ingenious Dr Sw—t

With Observations Historical, Political and Moral

Showing the derivation of this science from the Chaldees and Egyptians; with the particular practice of all nations on this important subject.

Also proposals at large for establishing a corporation for erecting 500 public offices of ease within the cities of London and Westminster for the conveniency of the nobility and gentry of both sexes in their natural necessities; which, besides the advantages that will thereby accrue to this noble metropolis, in particular will enable the company to divide above a million per annum to the great benefit also of every proprietor.

Dedicated to the profound Dr W—d,* and seriously recommended to all persons that drink the mineral waters of Pyrmont,* Bristol, Bath, Tunbridge, Epsom, Scarborough, Acton, Dulwich, Richmond, Islington, etc.

To the most learned

Doctor W—d

Profound Sir!

I have been a long time your humble admirer, but wanted an occasion to make the respect I bear you as public as your fame. The following essay has happily furnished me with the means, and though perhaps my manner of treating it may not deserve your regard, I am confident the world will unanimously agree with me in this: that the subject is truly worthy of your patronage. The prodigious discoveries you have made – and continue daily to make – in nature, and your known delight in subterraneous disquisitions, make me not in the least hesitate to invite you to descend with me for a quarter of an hour to the receptacles of a matter which was once an inhabitant of the human microcosm, and makes up – when discharged from our bodies – not a contemptible part of the elementary world.

To a philosopher there is no one thing more vile than another: his business is to be acquainted with all bodies, their compositions and properties, with the reasons of their changes. Whatever form matter is endued with, it is an object to him of contemplation, and the transformation of a pudding into a t—d merits no less to be considered than the growth of the corn of which the flour is made, which composes the main substance of the pudding. Nay, if any preference is given, it is rather a subject of so much more dignity, as the operations of nature in our bodies are of a higher estimation than those she performs in the earth, and as flesh and blood is of more value than dirt.

The house of office has ever been esteemed a place proper for sober reflection and study, but he must have a more than common turn of thought who makes the matter contained therein the subject of his contemplation.

How indefatigably you have laboured, and how vastly extended your knowledge in these parts, is unknown only to the ignorant and incurious. The learned world acknowledges universally your science in fossils – under which class, without any strain of etymology, t—ds may be reduced.

But for nothing are we more indebted to you, great sir, than for your recovery of that inestimable vase in which the divine Horace deposited his faecal burdens – which vase the silly vulgar are pleased to misname an urn. Oh, could you but in the same manner bestow on us some part of the treasure that pot once contained, what improvements might then be made in critical learning! The Roman lyric would then be perfectly understood, and B—tl—y* (if he can be ashamed) would blush at his comment.

I have now before me an ample field for panegyric – but I consider, sir, that you are only to be described by yourself in that happy style and phrase which everybody admires, but none can imitate. Accept then, sir, of this humble tender of my devotion to you, and the sciences you are so great a professor of, and permit me to subscribe myself,

Your disciple

and most obedient servant.

The Grand Mystery

The sun – the glory of the universe, the enlivening principle of animal and vegetable beings – instead of an object of profound admiration and enquiry, seems to an ignorant person but as a ball of fire, little bigger than an ordinary Cheshire cheese. Thus the moon shines, tides ebb and flow, the winds blow and seasons pass and return unobserved, at least unadmired, by the majority of mankind. Thus in short (to come to the subject of this essay) people of both sexes, of all ages, degrees, conditions, countries, complexions and religions, by night and by day, in sickness or in health, go to the sh–te, some in fields, some in houses, some in garrets, some in cellars, some in their beds and some in their breeches without the least reflection on the great and tremendous mysteries veiled under that performance, or imagining that their lives, fortunes and reputations depend on the regular and successful execution of it.

There is nothing the vulgar betray their ignorance and the wrong conceptions they entertain of things more in than when they bid a person, whom they would show their disesteem of, "Go shite", for can we wish our best friends a greater pleasure than to discharge those sensible membranes, the intestines, of a load which often produces such dreadful consequences when retained, and is always an occasion of fear to us till we are rid of it?

But this is not more in contempt of the person than of the action, the ill-judging vulgar always conceiving mean and unworthy ideas of such things as are most common in their sight, though perhaps they may at the same time be the things they are least acquainted with.

The sick, indeed, and they who purge for prevention, the ladies who take physic to preserve their shapes, and the beaus who sh—te for complexions, seem to have some imperfect regard to the dignity of this operation, and to entrust what they hold most dear to it. But if we examine into their views, we shall find them to be very superficial, they paying the chief honours, if their stools prove beneficial, to certain drugs or compositions of the apothecary's – which, without a proper disposition of the body to receive them, are but as a chip in the porridge. For there are few who conceive how much they stand obliged to the blood for throwing out the noxious humours into the bowels to the guts, which by their peristaltic motion drive them downwards, as also to the muscles of the anus for dilating and contracting to extrude them.

A man, to understand the whole process of the stercoral matter, besides being perfect in human anatomy, must be a profound philosopher, deeply learned in the doctrine of gravity and motion, and perfectly acquainted with the laws of statics. He must know, according to the old saying, "How many farts goes to an ounce" – a fart being only an eruption of merdal air, whereby the body it proceeds from is diminished in substance and weight. He must also reason upon and account for the variety we find in t—ds, their different consistency, colours and smells. He must know why my Lady Squitter does nothing but water, while country Jug leaves something at the bottom of a haystack as hard as a stone.

But these studies being the proper province of physicians and virtuosi, I would not offer to recommend them to the generality of my readers, and shall meddle with no part of the grand mystery of merdal exoneration but what relates, or may be of use, to us in our conduct of life: the regulations of our tempers, manners and constitutions of body and mind, and the improvement of our knowledge of ourselves and others.

It is very common, when we perceive the drift or design of a man, to say "We smell him", as if by some effluvia from his breech we come to the knowledge of the workings of his head. And indeed we have not a phrase in our language more significant or better founded than this, though the reason of it lies quite out of vulgar apprehension, it being certainly true that a person is never more effectually, more emphatically bewrayed (the old word for "betrayed") than when he or she is besh–t.

There are certain vapours which arise from the matter contained in the bowels, the particles of which, as they are variously figured, differently affect the brain and, operating on the judgement and appetites, produce in us the passions of hope, fear, joy, grief, love, ambition, etc., and are the efficient causes of our good or ill humour, our honesty or dishonesty to our neighbour, our opinions concerning ceremonies, etc., and our loyalty or disaffection to the state; and after the faecal matter is separated from the body, there is (while the excrement remains fresh) an exhalation of like particles which, ascending through the optic and olfactory nerves of any person standing over it, excite by sympathy the like affections in him, and inform him (if first duly instructed in these profound mysteries) of all that he can desire to know concerning the temper, thoughts – nay, actions and fortunes of the author of the excrement.

I hope therefore it will be no offence to my superiors that I propose, at the end of this little treatise, to lodge the supreme inspection of necessary houses in persons of more learning and better judgement than those who are now in possession of that office. The dignity of it is evident from the rods, the marks of honour and authority which even those ignorant fellows bear in their hands; but it will be in much higher esteem when occupied by philosophers and statesmen, who will be able – from the taste, smell, tincture and substance of the issue of our body's natural – to guess at the constitution of the body politic, and to inform and warn the government of all plots, designed revolutions and intestine grumblings of restless and aspiring men.

That there are ways of discovering these matters by a judicious survey of faecal dejections cannot be denied by anybody who is acquainted with history or knows anything of the transactions of our own times; such a one must needs have been informed how many conspiracies, intended rebellions and assassinations have been detected by a timely search of close stools and houses of office.

The ancient Italians and Romans were so well satisfied of the truth of this kind of divination that they never undertook anything of moment, either regarding the state or themselves as private men, before they had consulted the bowels of beasts for the event.

It was to this prevision the Roman Republic, in some measure, owed its success and grandeur; by this, likewise, that people lost their liberty, and the commonwealth became a monarchy; for I am confident that Caesar, who had been chief priest and was the most learned professor of enteroscopy in the world, would never have passed the Rubicon, nor undertaken that bold march against his country, if by his skill in this science he had not been assured the event would be happy.

Happy indeed had it been for him if his good fortune had not blinded him, and he had continued the use of such an inquisition, especially if according to the primitive usage and the practice of our modern revivers of the science, instead of the bowels of brutes he had consulted the dung of men. Had he commanded the whole Roman senate to sh–t before him (and who would have dared to refuse what Caesar commanded?), or had he only appointed trusty and intelligent visitors to inspect their several places of ease, he would have been sensibly forewarned of the dangerous melancholy of Brutus's temper, of the malignant humours that ran through Cassius's whole frame, and would have seen the figure of Casca's dagger in a t—d.*

I could produce from history innumerable instances of great men who, for want of a due regard to this science, have perished by the hands of vile murderers, but will content myself with one from our English chronicles.

Edmund Ironsides,* the most valiant of our kings before the Norman conquest – he who had so often triumphed over the invading Danes and had lately, in personal fight, got the better of their fierce king Canute – was basely killed by the villain Edric in a house of office as he was paying the necessary tribute to nature.* Too generous prince! It was not in him to suspect, least of all to look for, death in a place so homely, who had so often provoked it in the bloody field! He could not imagine danger could be there, and that a survey of the place was proper before he exposed those parts to an assault which could not see the assailant, else might he have discovered both the treason and the traitor, and might have caused him to be stifled in that scene of his villainy – a death suitable to a wretch, whose memory will stink as long as this fact shall be recorded.

I know when this treatise lights into the hands of sceptical persons and freethinkers they will treat it with ridicule, while others, though of orthodox faith in other respects, will regard it only as a piece of banter, having perhaps never before heard of soothsaying in a bog house.

Know then, ye men of wit and incredulous pretenders to it, that this science is as ancient as the Chaldees, by whom it was taught to the Egyptians, and by them transmitted to the old Italians and Greeks. But, as nothing religious or ceremonious is received so entirely by one nation from another but it meets with some alteration, as their customs and prevailing notions differ, we are not to wonder that in long time and far travelling this noble art of turd-conjuring became in part lost and so much changed as hardly to be known for the same.

That the Italians received it in its primitive purity I the rather believe, because of the propensity of that people to this day to be poking into human ordure; and that it degenerated by priestcraft, etc. – through the silly delicacy of the haruspices, whose weak stomachs puked at the sight or smell of man's dung – into an inspection of the entrails of brutes is plain from history. How it became afterwards quite laid aside in our parts of the world I could but dare not tell, for there are certain historical truths which it may not always be safe to relate.

Well fare then those great genii, those profound philosophers who to their immortal honour, with incredible labour and vast expense of time and thought, revived and restored to the world this inestimable mystery, this infallible method of knowing the inside of mankind and futurity. It is true it has hitherto, since its revival, been in the possession of few. Some great doctors and fellows of the Royal Society have been the only persons initiated – who, grudging that their fellow creatures should in anything be as wise as themselves, take great pains to make this science a secret. But as I, who am an adept, think myself under no restriction when the general benefit is in question, I hereby propose, if this meets with approbation, to publish a work I have by me, in the same volume, paper and letter as the *Constitutions of the Freemasons** was some time ago published, in which are laid down, in an easy and familiar method and style, all the principles and rules of this great science, whereby persons of the meanest capacities, old women and children will be instructed to find out the thoughts, actions, past or future fortunes, state of health and length of life of themselves or anybody else. And I hope I shall not fail of receiving from the learned world all suitable encouragement for so noble and useful an undertaking.

What emolument will the public receive by such a communication? What a new amusement will it be to those who are at a loss how to spend their time? And with what satisfaction will people go to sh–t when every stool shall be an improvement of their knowledge?

The gaiety of youth and gravity of age will be equally fitted, from hence, with matter of reflection or diversion. The midwife shall predict the child's fate from the first time it befouls itself, and nurses shall read learned lectures upon shitten clouts. The ladies will no longer consult their coffee and tea dishes for futurity. The oracles in Moorfields shall be struck dumb, and D–nc–n C–mpb–ll* shall be reduced to cry my works about the streets for his bread.

Then shall the desponding beau slide privately into his mistress's closet to search in her close stool whether she loves him or not; John the butler will see the form of his silver spoon in the housemaid's chamberpot, and suspected criminals shall be examined by trial ordural.

In this mirror of fortune, the smug chaplain shall see how long it will be ere he can mount to a bishopric. Here the noisy patriot may be fore-apprised whether his labours are like to be rewarded with a ribbon or a halter, and a furred gown and gold chain shall appear in every flying dragon the fated alderman shall throw out of his master's garret window.

By a recourse to this art shall all controversies be decided; lawsuits shall be no more, but Westminster Hall become desolate, when so easy and equitable a determination of rights may be had. A man's income shall then depend on his outgoings; his credit shall rise as they shall appear to fall; and the question shall be, when we would enquire a man's character, "How does he sh–t?"

In short, the time approaches when a chambermaid shall be a post of honour, and the groom of the stool, first M—r of St—, when bedmakers shall be dreamers of dreams and washerwomen shall see visions.

As for my part, I dare not arrogate to myself the least share of the honours which will be paid to these profound mysteries, but shall content myself to be known to posterity by the name of the *protocaccographer*.*

Till the aforesaid wonderful work is encouraged to appear, I forbear enlarging further on the mysterious part of sh–ting, but proceed to consider how far the practice of it, taken as a branch of private or civil economy, is capable of improvement, either in point of ceremony, decency, use or pleasure. As to the two first, I know very few who are masters of the grand air in sh–ting, the generality of people doing it with precipitancy and heat, as if frighted with what they are about; or else with indolence and unconcern, as if it were an action of no moment. The common form of letting down the breeches, the awkward postures in sitting, the frightful grimaces and barbarous exclamations now in vulgar use, all highly require a reformation.

I should therefore not think it amiss if academies were erected to be under the direction of persons of distinguished good breeding and ingenuity, where young gentlemen might learn to do what nobody can do for them, *en cavalier*, and little misses to sh–t in pots like ladies. They should be there taught how to walk to the house of office or close stool with a handsome air and step, and how to take up or let down their clothes in a genteel manner and to sit down with a good grace and in an inviting posture. They should there learn how to draw their features into agreeable forms and to utter musical and significant interjections. They should moreover be instructed in the art of wip—g – it being, as generally now practised, but what the Puritan calls the paint upon the face of the great whore, a filthy daubing.

In composing these forms, there ought to be a due regard had to the difference of sexes: that the exercise of gentlemen may be masculine and noble, and that the fair, when at stool, may – according to their nature – seem sweet and lovely. I would likewise have proper forms prepared for those who take physic or are troubled with gripes or piles.

And in order to prevent the confusion which often happens in these affairs, when time falls short, I would humbly propose that it may be the fashion for men to wear pulleys to their breeches and the ladies the like to their coats, and other undergarments; which should be so fitted with lines and weights that on loosing a cord in the twinkling of an eye, the male encumbrances may fall down to the heels and the female fly up to the waist.

The lower degrees of people will of course, as far as they are capable of imitating their betters, come into the fashion, and they who cannot afford a set of silver or ivory pulleys, double-gilt weights and silk lines will content themselves with twopenny cords and plain box and lead.

It will then be thought as necessary an accomplishment for the alehouse keeper's and tallow chandler's daughters to pass through the discipline of one of these academies as to go to the dancing school; and as footmen and chambermaids will learn of their masters and ladies, there will be such a circulation of politeness among the inferior orders of people as will totally wipe off that character of barbarity which is at present fixed by foreigners on our British vulgar.

I have likewise another proposal to make, which I submit to the consideration of a wise legislature.

London is now the largest city in the universe, and would be the most beautiful if its public buildings were answerable to the private, and our people were animated with a noble emulation to erect stately and convenient edifices for the common benefit. The Turks, as barbarous as they are here esteemed, outdo us by far in this, expressing all their magnificence in such buildings as are for everybody's use, such as caravanserais and baths.

It was this that made old Rome the glory of the world, that great city not being half so famous for its conquests and wide-extended empire as for its circuses, its theatres, amphitheatres and public baths.

In these last was to be seen all that art could perform with the most costly materials. They were built with the finest kinds of marble, and architecture and sculpture contended which should contribute most to their beauty. Here, besides bathing, a man might discharge his bowels with convenience and pleasure, the whole design of them being for common offices of ease and cleanliness.

Our climate, indeed, does not require – nay, does not allow – such frequent washings; but as we feed as plentifully as the Romans did, we have no less often occasion than they could have to discharge that part of our food which is of no use to our bodies. How useful would such places be to us? What honour would they do to this vast, this glorious metropolis?

There is nobody, I believe, who goes abroad but has been sometimes attacked in the streets by a sudden and violent motion to evacuate. What agonies are we then in? How disordered is our whole frame of body? And what care and dread sits on the countenance? The women fly to shops, where, after cheapening something they have no need to buy, and perhaps dropping the greatest part of their burden on the floor or into their shoes, they desire to speak with the maid; while we, unhappy wretches, hurry to some blind alehouse or coffee house where, before we can get a candle to light us to a nasty corner in the cellar, the fierce foe, too violent to be resisted, gains the breach and lodges itself on our shirts and breeches, to our utter confusion, sorrow and shame. And tho' they who keep coaches have the convenience of sh—ng under the seats, I believe they would be better satisfied to alight if, in every quarter of the town, there were handsome receptacles for them.

I despair of ever seeing such edifices erected at the charge of particular men, the noble spirit of the Romans and the charity of Mahometans* not dwelling among us; but when people shall be convinced, by the following scheme, that money laid out this way will be more profitably expended than in any other, I hope there will not be wanting among our wealthy gentry those who will undertake the execution of a project so conducive to their own and the general use and pleasure of the public.

Proposals

for

Erecting and Maintaining Public Offices of Ease
within the Cities and Suburbs of
London and Westminster

I. That a corporation be instituted by charter or otherwise as shall be thought fit, which shall be empowered to take in subscriptions for a sum not exceeding 25,000,000 *l.* sterling to be laid out in building five hundred sh–ting colleges, to be erected at convenient distances in the several parts of the town; the said corporation to be called "The Necessary Company", to be governed by a governor, deputy-governor and thirty directors, who shall be chosen every three years out of such members as have 10,000 *l.* stock. The stock to be transferable, as others are.

II. That the said colleges be built quadrangular, with portland stone, the porticoes and other ornaments in the front of marble; the statues, bas-reliefs and sculpture of the cornices and capitals of the pillars and pilasters being all designed to express some posture, branch or part of evacuation; the area to be paved with marble, with a basin and fountain in the middle, the group of which must likewise allude to that action; a covered walk with a flat roof, supported by columns, to run round the inner quadrangle, and between every two pillars a door to open to a sh–ting chamber or cell.

III. That the said cells be painted in fresco with proper grotesque figures and hieroglyphics; the seats to be covered with superfine cloth in wintertime, to be overlaid with Turkey carpets and in summer strewn with flowers and greens.

IV. That the men occupy the right-hand half of the square, and the other sex the cells on the left, from the grand entrance; and that the chambers on the female side be parted only breast high, for the better facilitating of conversation.

V. That each person, at his or her going away, leave with the waiter attending them the sum of two pence, to be applied to the benefit of the company, and divided quarterly or half-yearly, as at the general courts shall be agreed on.

VI. That all and every person be entitled, for the said two pence, to two pieces of clean, soft, whited-brown paper,* each piece containing eight inches in length and six inches in breadth. And whereas there are many studious people who have no leisure to read but while they are at stool, and who make a double use of books, first perusing diligently and then sacrificing the offspring of other men's brains to the issue of their guts, there shall in every college be a library from whence any person requiring it, instead of the blank paper aforesaid, shall be furnished with two pieces of B—t's legend* (the critical historian, etc.). The critic, or the humble admirer of the Muses, with some of Den—'s or A—e Ph—s's late easy labours;* the unfortunate lover shall be supplied from Dr Marten's or Dr Cam's treatises;* the Society of Cripplegate Grey Beards, with the late Father Jacob's pulpit declamations,* now ready to be put to the press for that purpose. And for the more elegant accommodation of people of quality and more than ordinary delicacy, there shall be an office in each of the said colleges to furnish them with gilt or India paper, velvet, satin, scarlet cloth, rabbit skins or other furs, or fine holland, they paying for the same according to the prices to be set by the court of directors.

VII. That the female attendants be matrons chosen for their volubility of tongue and knowledge in private history; and that the waiters on the other side be poor scholars or poets.

VIII. That each college be governed by a man of letters, a philosopher, whose salary, as well as those of the waiters, shall be settled by the directors of the company, and that there be likewise a cashier and accountant to each college.

IX. That all visitors by authority and virtuosi who shall bring a permit shall be suffered at any time to inspect the vaults of the colleges, and to continue in the same so long as they shall think fit.

X. The President, and other officers of each college, are to have lodgings over the cells.

XI. The great gates are to be open, and the waiters attending from five in the morning till eleven at night, and no person allowed to occupy a cell above half an hour, unless he or she is willing to pay proportionably for the extraordinary time.

XII. If any person having a genius for draught* shall presume to use his natural paint on the walls, such person shall forfeit five pounds, unless he chooses rather to lick off the offensive figures with his tongue.

There is room for a great many more improvements to be made upon this scheme. I shall therefore leave it, that other wits may exercise themselves upon it, and shall only say a word or two more to satisfy our moneyed men of the reality of the fund.

We may modestly suppose the number of persons sh–ting every day within the bills of mortality to be 1,200,000, of which I shall only take a third part, for those who by their circumstances will be enabled to frequent our places of ease. I likewise suppose that every person, at a medium, when in health, has occasion to go to stool twice a day (in sickness oftener). Now 400,000 groats, multiplied by the days in the year, will produce 1,433,333 *l.* 6 *s.* 8 *d.* The extraordinary income of the company, by engaging the physicians in its interests, the sale of goods in their vaults, the sole power of making saltpetre and distilling geneva* – which, perhaps, they may hereafter have interest enough to procure – these, I say, and other ways and means of increasing their revenue, may very well be computed to amount to half as much more; so that all charges defrayed, there will remain more to be divided on the capital stock than any of the present great companies can afford to give.

If the number of customers to those places is disputed with me, I only answer that I know the natural addiction of my countrymen to ease and pleasure so well that I am confident not only the nobility and gentry will stop up their private houses of office and repair hither, but even those whose fortunes are not very abundant will rather want one meal in three than not have the satisfaction of discharging the dregs of the other two in such delightful repositories.

The Wonderful Wonder of Wonders

Being an Accurate Description of the
Birth, Education, Manner of Living,
Religion, Politics, Learning, etc.
of Mine A–se.

By Dr Sw–ft

With a Preface and some few notes
explaining the most difficult passages.

Then as a jest for this time let is pass,
And he that likes it not, may kiss my a—

Jo Haynes

Preface

Gentle Reader,

Tho' I am not insensible how many thousand persons have been and still are with great dexterity handling this subject, and no less aware of what infinite reams of paper have been laid out upon it, however in my opinion no man living has touched it with greater nicety and more delicate turns than our reverend author. But because there is some intended obscurity in this relation, and curiosity, inquisitive of secrets, may possibly not enter into the bottom and depth of the subject, 'twas thought not improper to take off the veil and gain the reader's favour by enlarging his insight. *Ars enim non habet inimicum nisi ignorantem.** 'Tis well known that it has been the policy of all times to deliver down important subjects by emblem and riddle, and not to suffer the knowledge of truth to be derived to us in plain and simple terms, which are generally as soon forgot as conceived. For this reason, the heathen religion is mostly couched under mythology, and the Christian system in great part conveyed to us by parable. For the like reason (this being a fundamental in its kind) the author has thought fit to wrap up his treasure in clean linen and stow it in a bandbox, which it is our business to lay open and set in a due light; for I have observed, upon any accidental discovery, the least glimpse has given great diversion to the eager spectator – as many ladies could testify, were it proper or the case would admit.

But that the Reader may the better relish the contents of this subject, I advise him to keep a steady eye upon the single point in question, for the more his fancy roves, the more will the author's design 'scape his understanding; and give me leave to hope, in his name, that the nicest readers have no occasion to be disgusted from the familiarity of the subject, if they consider that the coarsest offices of life are indispensably necessary; if they consider, likewise, that the venerable scholiasts of antiquity, and even the Fathers of the Church, have not scrupled to dwell upon some grossities; and in poetry, Homer thought it no diminution to one of his heroes to have him caught turning his own spit, or to another to have him surprised in his nudities, tho' at the same time he put a young princess to the squeak.* Not has this subject furnished us with less wisdom than diversion, as the many proverbs in everybody's mouth – with reverence be it spoken – sufficiently testify. The politest companies have vouchsafed to smile at the bare name, and some people of fashion have been so little scrupulous of bringing it in play that, as our reverend author informed us in a late dissertation, it was the usual saying of a knight and a man of good breeding that "Whenever he rose, his a—se rose with him".

The Wonderful
Wonder of Wonders

There is a certain person lately arrived to this city whom it is very proper the world should be informed of. His character may perhaps be thought very incontinent, improbable and unnatural; however, I intend to draw it with the utmost regard to truth. This I am the better qualified to do, because he is a sort of dependant upon our family, and almost of the same age, tho' I cannot directly say I have ever seen him. He is a native of this country and hath lived long among us, but what appears wonderful, and hardly credible, was never seen before[1] by any mortal.

---

1. You must understand that the posteriors lie under an absolute necessity by their situation never to be seen *before*, but always – as the schoolmen term it – *ex parte post*.

It is true, indeed, he always chooses the lowest place in company, and contrives it so to keep out of sight. It is reported, however, that in his younger days he was frequently exposed to view, but always against his will, and was sure to smart for it.

As to his family, he came into the world a younger brother, being of six children, the fourth in order of birth[2] – of which the eldest is now head of the house, the second and third carry arms, but the two youngest are only footmen. Some indeed add that he hath likewise a twin brother who lives over against him and keeps a victualling house:[3] he has the reputation to be a close, griping, squeezing fellow, and that when his bags are full he is often needy – yet, when the fit takes him, as fast as he gets he lets it fly.

---

2. He alludes to the manner of our birth, the head and arms appearing before the posteriors, and the two feet, which he merrily calls the "footmen".

3. The belly, which receives and digests our nourishment.

When in office,[4] no one dischargeth himself or does his business better. He hath sometimes strained hard for an honest livelihood, and never got a bit till everybody else had done.

One practice appears very blameable in him: that every morning he privately frequents unclean houses, where any modest person would blush to be seen. And altho' this be generally known, yet the world, as censorious as it is, is so kind to overlook this infirmity in him. To deal impartially, it must be granted that he is too great a lover of himself, and very often consults his own ease[5] at the expense of his best friends. But this is one of his blind sides, and the best of men I fear are not without them.

---

4. Necessary house, which he afterwards calls "unclean houses".
5. This may be explained by the following ludicrous expressions: "Better out than in." "'Tis an ill tenant who pays no rent." "If these be your groans, the devil be your comforter." Etc.

He hath been constituted by the higher powers in the station of Receiver General, in which employment some have censured him for playing fast and loose. He is likewise Overseer of the Golden Mines,[6] which he daily inspects when his health will permit him.

He was long bred under a Master of Arts,[7] who instilled good principles in him, but these were soon corrupted. I know not whether this deserves mention – that he is so very capricious as to take it for an equal affront to talk either of kissing or kicking him, which hath occasioned a thousand quarrels. However, nobody was ever so great a sufferer for faults which he neither was nor possibly could be guilty of.

In his religion he has thus much of the Quaker that he stands always covered, even in

6. So called from the colour of the ore and the common term of gold-finders.*
7. Meaning the belly, in allusion to that passage of Persius: "*Magister artis, ingeniique largitor venter.*"*

the presence of the King; in most others points, a perfect idolater,[8] altho' he endeavours to conceal it; for he is known to offer daily sacrifices to certain subterraneous nymphs whom he worships in a humble posture, prone on his face and stripped stark naked, and so leaves his offerings behind him, which the priests[9] of those goddesses are careful enough to remove upon certain seasons, with the utmost privacy at midnight, and from thence maintain themselves and families. In all urgent necessities and pressures he applies himself to these deities, and sometimes even in the streets and highways, from an opinion that those powers have an influence in all places, altho' their peculiar residence is in caverns underground. Upon these

8. In leaving offerings at the necessary house, he alludes to the sacrifices offered by the Romans to the goddess Cloacina, president of all chapels of ease.

9. Gold-finders, who perform their office in the night-time. But our author further seems to have an eye to the custom of the heathen priests stealing the offerings in the night, of which see more in the story of Bel and the Dragon.*

occasions, the fairest ladies will not refuse to lend their hands to assist him: for altho' they are ashamed to have him seen in their company, or even so much as to hear him named, yet it is well known that he is one of their constant followers.

He lives from hand to mouth, but however the greatest and wisest people will trust him with all their ready money, which he was never known to embezzle[10] – except, very rarely, when he is sacrificing to his goddesses below.

In politics, he always submits to what is uppermost, but he peruses pamphlets on both sides with great impartiality, tho' seldom till everybody else hath done with them.

His learning is of a mixed kind, and he may properly be called a *helluo librorum*,* or another Jacobus de Voragine,* tho' his studies are

---

10. Too much haste on these pressing occasions, the author means, has often been the cause of dropping money out of our breeches.

chiefly confined to schoolmen, commentators and German divines, together with modern poetry and critics;[11] and he is an atomic philosopher, strongly maintaining a void in nature, which he seems to have fairly proved by many experiments.

I shall now proceed to describe some peculiar qualities which in several instances seem to distinguish this person from the common race of other mortals.

His grandfather was a member of the Rump Parliament,* as the grandson is of the present, where he often rises, sometimes grumbles, but never speaks. However, he lets nothing pass willingly but what is well digested.[12] His courage is indisputable, for he will take the boldest man alive by the nose.[13]

---

11. Here the author gives a clean wipe on these performances as most contemptible in themselves – consequently most proper for the necessary house.

12. The beauty of this expression lies in the ambiguity betwixt the digestion of thought and food.

13. This is explained by the note no. 5.

He is generally the first a-bed[14] in the family and the last up, which is to be lamented, because when he happens to rise before the rest, it hath been thought to forebode some good fortune to his superiors.

As wisdom is acquired by age, so by every new wrinkle[15] in his face he is reported to gain some new knowledge.

In him we may observe the true effects and consequences of tyranny in a state: for as he is a great oppressor of all below him, so there is nobody more oppressed by those above him. Yet in his time he hath been so highly in favour that many illustrious persons have been entirely indebted to him for their preferments.[16]

---

14. This relates to the proverb "You rose with your a–se foremost, you are so lucky today".

15. This refers to another: "You have one wrinkle in your a–se more than you had before".

16. I refer the Reader for an explanation of this passage to Bembo's *Lives of the Cardinals*.*

He hath discovered from his own experience the true point wherein all human actions, projects and designs do chiefly terminate,[17] and how mean and sordid they are at the bottom.

It behoves the public to keep him quiet, for his frequent murmurs are a certain sign of intestine tumults.

No philosopher ever lamented more the luxury for which these nations are so justly taxed; it hath been known to cost him tears of blood.[18] For in his own nature he is far from being profuse, tho' indeed he never stays a night at a gentleman's house without leaving something behind him.

He receives with great submission whatever his patrons think fit to give him, and when they

---

17. I conceive the author means no more by this than that our highest performances, either of hand or head, in plain words, amount to no more than a t—.

18. Haemorrhoids, according to the physicians, are a frequent consequence of intemperance, which is here our author's meaning.

lay heavy burdens upon him, which is frequent enough, he gets rid of them as soon as he can, but not without some labour and much grumbling.

He is a perpetual hanger-on, yet nobody knows how to be without him. He patiently suffers himself to be kept under, but loves to be well used, and in that case will sacrifice his vitals to give you ease; and he has hardly one acquaintance for whom he hath not been bound[19] – yet, as far as we can find, was never known to lose anything by it.

He is observed to be very unquiet[20] in the company of a Frenchman in new clothes or a young coquette.

He is, in short, the subject of much mirth and raillery, which he seems to take well enough, tho' it hath not been observed that ever any good thing came from himself.

---

19. This turn of humour depends on the different effects of being *bound* in law and *bound* in body.
20. Their tails being generally observed to be most restless.

There is so general an opinion of his justice that sometimes very hard cases are left to his decision – and while he sits upon them, he carries himself exactly even between both sides, except where some knotty point arises; and then he is observed to lean a little to the right or left as the matter inclines him, but his reasons for it are so manifest and convincing that every man approves them.

The Blunderful Blunder of Blunders

Being an Answer to the

Wonderful Wonder of Wonders

By Dr Sw—ft

Ars longa.

The Blunderful
Blunder of Blunders

Having lately perused a paper called *The Wonderful Wonder of Wonders*, I could not but with the highest resentment animadvert upon its author, who at this time of day, when all heads are at work about affairs of the greatest consequence, should be so cruel as to write upon a plain subject with so much obscurity, whereas the naked truth always appears best in a simplicity of language.

'Tis true indeed that in the early ages of learning scholars, through an affected vanity of appearing wiser than the rest of mankind, disguised their knowledge under the cover of hieroglyphics, but as philosophy acquired more heat and lustre, these clouds began to vanish, and the rays of truth more universally diffused themselves to all such as was earnest to search and pry into the secrets of nature. So were the oracles of old, by direction of their father, the Devil, wrapped up in the utmost darkness, for he would have them carry his own black stamp till men of penetration and judgement discovered their fallacy and rescued the deluded reason of their admirers from its greatest enemy, ambiguity.

Now I appeal to all men of reading or experience, whether it has not been the constant practice of impostors, the better to carry on their cheats, ever to amuse the world with riddles. Which puts me in mind of a story that will not be amiss in this place. A certain cunning fellow who had been reduced to his last shifts, and knowing the world to be very fond of wonders, gave out that he had found a terrible monster in a wood, which he took care to chain up in a dark corner of his room. People flocked in abundance, and the man made a very great advantage of his show, for he managed it so dextrously, by the dreadful accounts which he gave of its fierceness, that nobody durst approach near enough to see what it was, till one day a pot-valiant fellow, who knew how to value his sixpence, rushes upon the monster, swore he would see what he was to have for his money and, in short, drags out a dog in a doublet.

The Reader may expect I should say something here, but I ask his pardon if I refer him to the conclusion for an application.

I would willingly expostulate with my friend, and ask him what would he think if Nature, in her works, should proceed in a method enigmatical, that every species of fruit should have a dark skin drawn over it, insomuch that we should not be able to distinguish between an apple and a peach, between a pear and a nectarine, without stripping them of their clothes, would not a vast number of inconveniencies ensue? Or should she thro' whim and frolic affect eclipses in sun, moon and stars, the world would have a fine time on't. Should every lady run into the frolic of glueing their masks to their faces, there would be an end of beauty. In short the evils of disguise are without number – for which reason, truth is dignified with the epithet of "naked", and our author, without ceremony, should have uncovered his subject to the world: it was not of so little importance to mankind as to be concealed from them.

When a man writes, either for the information or improvement of the world, let him write to be understood by the world. The reason I insist upon this so much is – I was in company the other night with six gentlemen of as good understanding as any in Ireland, and without vanity I may say as any in England, where this same paper of "wonderful wonders" was introduced – ay, and read over three times before anyone durst venture even at a conjecture. "At last we began to debate it," says one. "I fancy this must be a kind of a satire upon Jo. D—r* because he is described as 'a close, griping, squeezing fellow'." "No sir, that cannot be," said another, "for you know he is made to say, 'as fast as he gets, he lets fly' – besides, the man is dead." Said a third: "I have it – depend upon this – that it is meant of a judge, because he is a great oppressor of all below him, and you know he is given to frequent murmurs." "Not at all," said I. "That cannot be, for there is no judge in town observed to lean either to the right or left. Indeed, were it not for that, I should be inclinable to think so, because this person is described to peruse pamphlets on both sides with great impartiality." Upon this conviction I acquiesced, and a friend next me rises with some appearance of reason, and said the presumption was strong of his side that it must be a b—p,* because his studies were confined to schoolmen, commentators and German divines. But this was soon overthrown, because no b—p has any civil employment, and the person here mentioned is made Receiver General.

A certain grey-headed reverend divine in town said he was sure it was meant of the Wooden Man in Essex Street.* Now I humbly beg leave to start these queries to him:

Query – Whether the Wooden Man in Essex Street ever goes to bed.

Query – Whether he ever leaves anything at any gentleman's house.

Query – Whether he be lately arrived to this city: it is well known that he is an old stander, and one of the ancient inhabitants of it.

Query – Whether he was never seen before by any mortal.

Query – Whether he frequents unclean houses, at least in the plural number.

Query – Whether people trust him with their ready money.

Query – Whether his grandfather was a member of the Rump Parliament.

Query – Whether he ever sheds tears of blood.

Now, *Quære* – Whether the Wooden Man would not have guessed as well.

One held it to be a Jacobite paper, and that he saw the Pretender* at the bottom, under the name of Jacobus de Voragine.

But to be short, after many long arguments and debates, one in company (*non quia nasus cœteris nullus erat*)* started up and said, "Gentlemen, I smell a rat – it is my arse all over..." And we all applauded his penetration.

Postscript

If the gentleman – thro' consideration of the losses sustained by the South Sea* – has, out of a design to encourage trade and commerce, sold the public a bargain, I heartily ask his pardon for these animadversions. But if not, he may expect much severer in my next, together with an ample dedication to the gold-finders of the city of Dublin.

NB: The author of this answer intends very soon to oblige the world with a historical account of bargains.

Serious and Cleanly Meditations

upon a House of Office

Dedicated to the
Gold-finders* of Great Britain

By Cato

To which is added

The Bog House, a Poem
in Imitation of Milton

To the

Gold-finders

of Great Britain

Gentlemen,

I have long and often wondered that amongst all those different bodies of men who have had their names illustrated in modern dedications, so useful a society as yours should want its due encomiums, but – alas! – corruption is become public and universal, which beginning amongst the great ones takes its course downwards and infects the vulgar herd. Our mercenary scribblers, in imitation of their depraved patrons, make lucre the chief end of their writings and, finding more advantage in glossing over vice than in doing justice to virtue, prostitute their pens to the most shameful purposes. Yet indeed I cannot blame their conduct in this particular, for if they expect extravagant prices for their panegyrics, to whom can they address them so well as to those who least deserve them?

You cannot be ignorant, gentlemen, that I have been long engaged in the service of my distressed country in exposing the tools of power and the creatures of ambition; that I have laboured day and night in this glorious (but neglected) cause, and endeavoured to put Church and State upon such a foundation that the latter may appear unexceptionable (even to myself) and the former without spot or wrinkle, or any such thing. In pursuit of these labours, and as opportunities have offered themselves, I have made it my business (as it is the greatest pleasure of my soul) to point out real merit to the world; you cannot forget that I have already addressed myself, in three immortal pieces, to a noble lord, the Man in the Moon and the clergy of the Church of England; and after these, I think I cannot make a better choice of my patrons to this piece than the gold-finders of Great Britain.

Nor do I think that this address comes improperly from my pen, since there seems to be so great an analogy between our professions, which both consist in carrying away the inward corruptions of mankind; and there is yet this further likeness: that altho' our labours tend alike to sweetness and decency, yet we make a great stink in the operation. I confess the advantage seems to lie on your side, in the eye of the world, by confining your labours to the night, when people are asleep, and out of the smell of them, whereas I perform publicly at noonday. But I shall never be ashamed to give public offence in order to do public good – which ever was, and ever shall be, the avowed end of my writings.

It is often objected to us both in derision that we get our bread by raking into ordure and nastiness. I wonder how any men can be so hardy as to urge a reflection against us which is stronger against some persons in authority, who are well known to enrich themselves and their families by dirtier jobs than any which either of us are employed in.

When – alas! – will the labours of reformation and gold-finding be at an end? When will mankind cease to sh–te and be corrupt? Alas, never whilst they continue to be men, for it is natural to them, and interwoven with their constitution. Why therefore should we upbraid human creatures for what is human and common to ourselves? Why rather do I upbraid them? For you indeed do not, but do your work quietly, and receive your money thankfully, without stigmatizing those who find you employment, whilst I am eternally snarling at mankind for what is essential to mankind, and abusing those human frailties by which I get my bread. Ridiculous beyond anything which I have ridiculed to blame the bridge which I go over, and forget the old saying that "Without sins there would be no taxes"!... If all the world were virtuous, how would Cato differ from other men?

The rest of my thoughts on this important subject are contained in the following meditations, which I beg leave to make public under the protection of your most illustrious society, as a mark of that affection and sincerity with which

I am,

    Gentlemen,

        Your constant fellow labourer

            for the good of our country,

                – Cato

Serious and Cleanly Meditations

As I was one morning sitting at my ease in a pensive manner in one of the public offices in this great city, I fell into a train of meditations suitable to the dignity of the place and the awfulness of that solemn occasion.

I could not forbear reflecting on the vanity of human pride, which puts us to such immense cost and trouble in tricking out a tawdry outside with velvet, brocade, fine laces and flaxen wigs; whilst if we look into ourselves and consider well the furniture of our hidden apartments, we should have little reason to boast of the beauty or comeliness of our persons. I never enter into the drawing room, the playhouse or any other public assemblies of the beau monde without laughing in my sleeve at the various fopperies of both sexes, which I look upon as so many artifices and cover-sluts* to hide their inward deformities; for methinks it is enough to mortify the proudest of us all to remember that although we may adorn our outward parts with fine clothes and sweeten them with costly essences, yet our insides are no better than common bogs, and are composed of materials too foul to name. It is no doubt, upon this account, that we are said by St Paul to be made of the filth of the world, and the offscouring of all things unto this day. For the same reason, I was mightily pleased with the reply which a philosopher made upon hearing a certain lady very much extolled for her beauty: "Ay," said he, "she's a very fine creature, but for all that, she stinks when she goes to stool."

This is a true state of the human body, and I fear that, upon examination, our minds will appear but little better, and that the emanations of one do not much surpass in cleanliness the voidings of the other. If we survey the learned world, as it is called, what do its modern productions consist in but right modest *Cases of Impotence*, *Essays upon Farting* and the *Wonderful Wonders of Mine A–se*? What is thy shop, O C—l,* but a bog house, the common sink of lewdness and bestiality? Is it not equally avoided, unless in cases of necessity, by all persons of modesty and good breeding? And is it not as much frequented by those who, like swine, delight to wallow in mire and nastiness?

Secondly, bending my eye downwards into this subterraneous cavity, I said to myself: "Does man live for this? Do all his pursuits tend only to increase these stenches and swell this noisome profundity? Alas, for nothing else! The toils and ambition of the great, as well as the labours and fatigues of the vulgar are subservient to this end. For what do we live but to eat and drink, and exonerate ourselves in these voracious abysses? The body of man is but a thoroughfare to these common receptacles of all things. What an infinite variety of creatures is here blended together? Methinks I see (and oh, that I could not say I smell) ten thousand various dishes tossed up together and jumbled into a second chaos of matter."

Amongst this compost or mess of medley, I spied the leg of a human child sprawling up in a perpendicular manner, the sight of which drew from me several pathetic exclamations against the heinous and crying sin of fornication – a sin which is the mother of a much greater one: the sin of murder, even the unnatural murder of our own offspring. "Alas!" said I. "How many innocent babes have been sacrificed to these infernal deities and buried as ignominiously as they were begotten?" Hereupon I could not but admire the wise institution of the Scotch nation, which, in order to prevent a crime of so deep a dye, will not suffer any such destructive chasms within the limits of their kingdom.

Thirdly, bending my eyes down again into the same odoriferous gulf, I beheld it enriched on every side, with scattered reams of ancient and modern learning, with the excrements of the head as well as the tail. Here lay a passage in *Tom Thumb*, and there a sentence in St Austin;* here a scrap of a ballad, and there a fragment of metaphysics; while dispersed up and down in a confused manner, I observed the following title pages, half-obliterated with the rising vapours of the place, viz. *Duns Scoti Opera Omnia – Manual of Devotions – The Aeneid of Virgil, translated by Joseph Trap, A.M. – The Several Protests, etc. – The Dangerous and Sinful Practice of Inoculation – The Scourge – The Life and Adventures of Robinson Crusoe – Letters to the Late Earl of Sunderland, by Archibald – A New Almanack for the Year 1713 – The Half-penny Post – The Weekly Journal* –*

and, amongst the rest, some of my own laborious and (as I vainly hoped) immortal lucubrations.

This melancholy occurrence threw me into a reflection to which I was before a stranger, and first taught me – whilst I so liberally bestow contempt on other authors – how ill I can bear contempt myself. With what calmness and equanimity did I behold the miscellaneous remnants of my brother writers lie rotting in this infamous obscurity! But when I spied the least part of my own dear issue, how did it damp my fond elated heart, and fully in my own eyes the glory of my late productions! How little did I expect to see this day! Must the great Cato die then like common scribblers and perish in the same foul mass with his ignorant antagonists? Shall it be said in future ages that mist and Cato underwent the same fate – that after infinite wranglings and mutual altercations they were at last profaned by the base hands of porters and chairmen, and retired together to this stinking asylum? It must be so – O Pope! Thou reason'st well:

> Whom not th' extended Albion could contain,
> From old Belerium to the northern main,
> The grave unites – where ev'n the great find rest,
> And blended lie th' oppressor and th' oppressed!*

Oh! that I had never swerved from my first honest principles, nor exposed myself by daily inconsistencies and self-contradictions! That I had not deferred the cause of the Protestant succession to gain the poor popularity of pleasing a Jacobite rabble, nor made myself the wretched tool of one who, by his pride, avarice, ill nature and implacable malice against all persons above him has made himself contemptible in the eyes of the world! Better had it been for me to become that detested thing, a court parasite, than the servile instrument of an ill-mannered country squire, of a noisy coffee-house politician – now art thou Cato indeed, and hast committed a far greater murder on thyself: he only killed a perishable body, but thou hast stabbed a much dearer part, a flourishing reputation, and laid violent hands on thy own excellent arguments.

Whilst I was thus lamenting over my deceased offspring, which lay buried in this ignominious grave, and wondering at the ignorance of mankind, which confounds in this undistinguishing manner the labours of the learned and unlearned together, I cast my eyes round the walls of the venerable edifice, and was not a little relieved from these uneasy thoughts by the various flights of poetry which I found inscribed there. It was some consolation to my afflicted soul to find that the same place which every day swallows up some part of learning inspires its inhabitants to supply the world with others in its room – and that what seems the open sepulchre of the belles-lettres is also the nursery of wit and the seat of the Muses.

I know it would be an agreeable entertainment to the public to present them with some of those little mural pieces which I perused there; but I am afraid that it may seem unworthy the character I bear in the world (though they are all excellent in their way), as well as inconsistent with that gravity and seriousness which I proposed to myself in the course of these learned meditations. However, I will indulge the impatience of my Reader with two or three, which may serve for a specimen of the rest, and contain in them sharp strokes of wit, as well as very useful morals.

The first I shall mention is a philosophical one, and seems, by the reasoning in it, to be written by that profound author, the famous Phileleutherus Lipsiensis,* and is as follows:

The learned hold in ev'ry nation
Four methods of evacuation,
Videlicet: *imprimis*, writing;
Secundo, pissing; *tertio*, sh—ting;
Quarto – but that is better guessed,
By far the shortest and the best;
Of all which methods there proceed
Three from the tail, and one the head;
From whence 'tis easy to explain
Why the tail's stronger than the brain.

The next is written in the true spirit of heroic loyalty, and claims for its author that renowned knight, Sir James Baker;* observe, Reader, how zealously he writes for his King and country.

The man who to his country is a friend
May downy peace and gentle stools attend;
But him who would our free-born rights destroy
May costiveness and racking piles annoy.

The author of the following one is uncertain, but being written in the Pindaric style, it is attributed by most critics either to Mr Tho. D—f—y, or Mr A—n H–ll.*

Such places as these
Were made for the ease
Of every fellow in common;
But, to poets who write
On the wall as they sh—te,
'Tis a pleasure far greater than woman;
For he's eased in his body and pleased in his mind
Who leaves both a t—d and some verses behind.

For the encouragement of these little geniuses, I condescended to unbend myself in poetry, and wrote the following lines on a clean part of the wall:

By the British Cato

If liberty, as I have taught, be good,
'Tis surely best in this place understood:
Here we may sh—te, fart, stink, do what we please –
This place was made for liberty and ease.

I am glad to find there are some poets in the world who know their own strength and do not endeavour to soar further than their abilities will bear them – I hope others will follow so laudable an example and reduce their studies to a better standard; several bards (whom, to avoid ill-will, I'll forbear to name) have given themselves and the world abundance of trouble to no purpose, in epic and dramatic writings, who might do their country a great deal of honour, and themselves a great deal of service, if they could be prevailed upon to confine themselves to these Sir-Reverential performances: but as Horace says:

*Omnibus hoc vitium est cantoribus atque poetis**

There is no convincing a headstrong poet of his errors, and therefore 'tis in vain to throw away good advice upon such impenetrable ingenious blockheads: let them write on and be damned to this infernal dungeon for ever.

I then fell into a philosophical argument with myself, to explore the reason why these houses of ease are always so productive of poetry beyond what we observe in other domes; and first of all I thought it not impossible that the miscellaneous poems which crowd these walls might be the *manes* or ghosts of those unhappy rhymes which lie interred underneath – which, being restless in that condition, wander up and down these places, in the nature of other spirits. But I was soon driven out of this opinion by two considerations. First, because these poetical excursions are generally apposite to the place upon which they are inscribed, and are very often sarcastical repartees on one another. Secondly, because they are observed to appear by day as well as by night – which, I think, is allowed by all divines to be inconsistent with the nature of apparitions.

Being therefore obliged to quit this conjecture, I had recourse to another solution, which I think much more probable. I suppose that the poetical matter, which lies in these dark caverns, throws up certain fumes which, finding an easy passage through the body of him who sits brooding over them, ascend into the brain and there create an inclination to versifying; under which condition, and in the luxuriancy of his vein, he transcribes his thoughts out of his head upon these walls in imitation of the ancient poets whom Horace mentions.

This I think a tolerable hypothesis, and is confirmed by a parallel instance: that of churchyards and charnel houses. It is generally allowed, if any person sits for a considerable time by himself on a tombstone or a coffin, that he is imperceptibly affected with gloomy and melancholy ideas; this could proceed from nothing but those unwholesome vapours which arise from rotten and putrid carcasses. And if it be granted that such carcasses can inspire us with suitable thoughts, I desire to know why the more volatile remains of dead rhymes may not produce the contrary effect and generate a sudden appetite to rhyming. I am further convinced of this by a couplet I found here, which I read with a great deal of pleasure, because I do not love to be singular in any of my opinions; it was this:

> If smell of t—d makes wit to flow,
> L—d, what would eating of it do!

Yet after all, however well grounded this conjecture might be, I am resolved not to be so opinionated of my own judgement as absolutely to determine this knotty point, but shall leave it to the more accurate decision of the Royal Society and other literati of the nation. I fell into several other reflections on this occasion, which I have either forgot or do not think proper to trouble the Reader with. I also made several discoveries, which the niceness of my honour will not allow me to communicate, it being my opinion that everything which is dedicated to these infernal abysses ought to remain in everlasting oblivion.

I will conclude all with this instructive moral: that however we may value ourselves for our superiority over one another, there is no real difference between man and man – especially upon these necessary occasions, when (according to the proverb) "a beggar is as great as a king".

The Bog House

A Poem

Humbly Inscribed
to the Late Translator of Virgil

Of man's important bus'ness, and his work
Of nature, late and early, ev'ry day,
Sing, my Pierian Muse, in numbers sweet
As is my subject, voiding all thy wit
Uncostive, flowing forth in happiest strains.
The swain surcharged with plentiful repast,
Or rural banquet, or domestic meal,
Whether at morn, when Turkish berry adust*
(Fell enemy to sleep, and cause of spleen)
Or Indian leaf suffused with fragrance bland
Comforts the maw, or solid oatmeal food –
Hight "hasty pudding" –
Which heats the blood of Caledonian swains
And warms the north, or roast and boiled at noon,
Or well-sauced herbage, with cold lamb at ev'n,
Full fraught retires. To house unpraised on mount
He hies, vile eminence, convenient site
For work unsavoury, or to garden side,
Where breath of May and odoriferous flowers
Do qualify with sweets th' offensive scent,
Or where cool rivulet with limpid stream,
Running fast by precipitates the filth,
Purging the dome: Alcides so of old
Th' Augean stable rinsed* – thither repairs
The loaded swain incontinent to pay
Tribute of ordure to the gods of earth,

— 93 —

Brethren of Moloch.* Not the eastern shrine
Frequented more in Araby the blessed,
Where uplift prophet, dubious, hangs in air,
The strife of magnets. In the dome appears
A graduated seat, for infant bum
Or veteran, built of tree, Norwegian spoil,
Or such as Danzig yields, the prince of woods.
Here triple hole discovers hollow womb
Of earth, dreadful to sight, abhorred to smell.
Up from the putid dungeon fumes ingrate
Ascending hurt the sense; careful to ken,
Lest hapless he may light on foul remains
Of dirty clown, with galligaskins loosed
(Oft fatal to the purse, or watch in fob,
Which well rewards gold-finders' filthy toil),
Bending oblique, his postern he applies
To perforated board, as erst were wont
Apollo's votaries to submit their ears
To Delphic tripod, and receive response
From pagan shrine: so here from end reverse,
Sounds are immitted,* to invoke the sprites
Of darkness and alarm the powers of night.
And now the swain at ease, composed to vent
Embowell'd food, from Nature's secrets stores
Discharges plenteously of every kind:
Corn, fish and fowl, and wine of various taste,
Caecuban or Falern. What earth, sea, air affords
(Vile refuse of concocted aliment)

With bountiful effusion is bestowed:
Bursting it flies, conveyed by force of wind,
And tremulous noise, sent downwards all at once,
With horrid violence, like Etna's wild
Eruption and the fall of craggy rocks
Inwards on Mount Vesuvius, or Nile
Spewing with all his mouths into the sea,
Or sulph'rous vapours, kindled in the air
With nitre, conflict of elements – so roars
The darksome cell, with repercussive sound
Of postern gun disploded, which reports
Afar and echoes from the vast abyss.
Thus he, thrice happy, in luxuriant stools
Voids the successive gathering of his meals.
As when a bee, with balmy juice replete
And liquid spoils of gardens taken short,
Flies hastily to waxen privy house
In hive or hollowed oak, or chimney top
(Besmeared with soot, of taste contrariant),
Or ruinous wall, and laxative refunds
(Sweet voidance) all the bev'rage of the day.
Such blessings Heaven ever has denied
To sinful mortals, when astringent food,
Or body-binding claret, bars the port –
Painful coercion, cause of inward heats
And fierce distortions of the face and trunk –
The Gods vouchsafe me gentle stools and ale!
Meantime a soft abstersive* is prepared

By foliage of fields or books supplied,
Or verdurous plant, good herb, or pliant dock,
Delusive if unfolded, or trefoil,
Jerne's* vegetable pride, or hay,
Fodder of man and beast, or above all
Love-laboured sonnet, or some senseless rhyme
On the disdainful nymph, or poor conceit
Of paltry scribbler, starvèd as himself,
Be it in verse or prose, or smart lampoon
On Church and State – all read with profit here –
Or Dutchman's commentary, long and dull,
Or ven'mous work of critic, or divine
Polemic – nor might Jove himself disdain
S—'s* foul paper at celestial stools.
These strew the place and fill the mural void,
And claim the fundamental office to wipe bum –
Wonderful bum! – subject of modern wit
And hidden cause, for to thy secret power
And kindly operations, Nature owes
Motions of wit and mirth and joyous thoughts.
When thou art open, fancy flows apace,
But when retentive, merriment's entranced
In spleen, and lies in clouded brain
Incarcerated: medicinal wood,*
Thy porter, opes and shuts thy folding doors,
Still kind to me, propitious to my verse.

The Bog House

and

Glass-window Miscellany

Taken from Original Manuscripts,
Written in Diamond, etc.

by Some Persons of the First Rank
in Great Britain.

The first epigram, being philosophical, is supposed to be written by that profound author, the late Dr Ch—ne of Bath.*

> The learned hold in ev'ry nation
> Four methods of evacuation,
> Videlicet: *imprimis*, writing;
> *Secundo*, pissing; *tertio*, sh—ting;
> *Quarto* – but that is better guessed,
> By far the shortest and the best;
> Of all which methods there proceed
> Three from the tail, and one the head;
> From whence 'tis easy to explain
> Why the tail's stronger than the brain.

The next is written in the true spirit of heroic patriotism, and claims for its author the Right Hon. B—b D—n, Esq.*

The man who to his country is a friend
May downy peace and gentle stools attend;
But him who would our free-born rights destroy
May costiveness and racking piles annoy.

The author of the following one is uncertain, but being wrote in the Pindaric style, it is attributed by most critics to Mr A——n H—ll, or the Laureate.*

Such places as these
Were made for the ease
Of every fellow in common;
But, to poets who write
On the wall as they sh—te,
'Tis a pleasure far greater than woman;
For he's eased in his body and pleased in his mind
Who leaves both a t—d and some verses behind.

For the encouragement of the little geniuses, I condescended to unbend myself in poetry, and wrote the following lines on a clean part of the wall:

If liberty, as I have taught, be good,
'Tis surely best in this place understood:
Here we may sh–te, fart, stink, do what we please –
This place was made for liberty and ease.

The following is supposed to be wrote by the late
Bishop of ——:

If death doth come as soon as breath departs,
Then he must often die who often farts:
And if to die be but to lose one's breath,
Then death's a fart – and so a fart for death.

The following was wrote by Capt. B—ns,* in a
bog house at Epsom Wells:

Privies are now receptacles of wit,
And every fool that hither comes to shit
Affects to write what other fools have writ.

In the Rose Tavern bog house, wrote by Mr T.C.
the player:*

D—n their doublets, and confound their breeches –
There's none beshits the wall but sons of b—ches.
May the French pox and the devil take them all
Who beshit their fingers, then wipe them on the wall.

At Kensington, wrote by a beau who hated tobacco:

This is a place that's very fitting
To piss, to fart, to smoke and shit in.

In Oxford, wrote by Mr Archdeacon:

With such violent rage
Sir John did engage
With the damsel which he laid his leg on
That his squire who stood near
Swore it looked like a spear
Of St George in the mouth of the dragon.

At St J—'s, humbly addressed to his M—y:

The greatest monarch, when a-fighting,
Looks not so fierce as G— when shiting.

Wrote in a bog house which stands over the water, at the Spread Eagle in Bunny, in Nottinghamshire:

> The nicest maid, with the whitest rump,
> May sit and shite, and hear it plump.

Wrote in a bog house at the Bush in Carlisle, in Cumberland.

> Within this place two ways I've been delighted:
> For here I've swinked, and here I've shited.
> They both are healthful, nature's ease require 'em,
> And tho' you grin, I fancy you desire 'em.

Underwritten:

> What beast alive could bear to swink
> In such a filthy hole as this is?
> The nauseous stink might, one would think,
> Disturb his taste for am'rous kisses.

Underwritten:

This was wrote by some beau,
The fop you may know,
His squeamish exception would make one believe it;
Tho' the smell where we shit
Is not grateful a bit,
Yet I ne'er knew a c—y* that savoured of civet.

On a fart:

Gentlest blast of ill concoction,
Reverse of high-ascending belch;
The only stink abhorred by Scotchmen,
Beloved and practised by the Welsh.

Softest note of inward griping,
Sir Reverence's finest part;
So fine it needs no pains of wiping,
Except it be a brewer's fart.

Swiftest ease of colic pains,
Vapour from a secret stench
That's rattled by the unbred swains,
But whispered by the bashful wench.

Shapeless fart, we ne'er can show thee,
But in that noble female sport
In which by burning blue we know thee,
Th' amusement of the maids at court.

On a fart, let in the House of Commons:

Reader, I was born, and cried;
I cracked, I smelt, and so I died.
Like Julius Caesar's was my death,
Who in the Senate lost his breath,
Much alike entombed does lie
The noble Romulus and I;
And when I died, like Flora fair,
I left the Commonwealth my heir.

The Wonder of all the Wonders that ever the World Wondered at

By the Author of the
Art of Punning,
Benefit of Farting, etc.

Quæ majora putes miracula?[*]

The Wonder of All the Wonders

that ever

the World Wondered at

For all Persons of Quality, and Others

Newly arrived at this city, the famous artist John Emanuel Schoitz – who, to the great surprise and satisfaction of all the spectators, does the following wonderful performances, the like before never seen in this kingdom.

He heats a bar of iron red-hot and thrusts it into a barrel of gunpowder before all the company, and yet it shall not take fire.

He lets any gentleman charge a blunderbuss with the same gunpowder and twelve leaden bullets – which blunderbuss the said artist discharges full in the face of the said company without the least hurt, the bullets sticking in the wall behind them.

He takes any gentleman's own sword and runs it through the said gentleman's body, so that the point appears bloody at the back to all spectators; then he takes out the sword, wipes it clean and returns it to the owner, who receives no manner of hurt.

He takes a pot of scalding oil and throws it by great ladlefuls directly at the ladies without spoiling their clothes or burning their skins.

He takes any person of quality's child, from two years old to six, and lets the child's own father or mother take a pike in their hands; then the artist takes the child in his arms and tosses it upon the point of the pike, where it sticks, to the great satisfaction of all spectators, and is then taken off without so much as a hole in his coat.

He mounts upon a scaffold, just over the spectators, and from thence throws down a great quantity of large tiles and stones, which fall like so many pillows without so much as discomposing either perukes or headdresses.

He takes any person of quality up to the said scaffold, which person pulls off his shoes and leaps nine foot directly down on a board prepared on purpose, full of sharp spikes six inches long, without hurting his feet or damaging his stockings.

He places the said board on a chair, upon which a lady sits down, with another lady in her lap, while the spikes, instead of entering into the under lady's flesh, will feel like a velvet cushion.

He takes any person of quality's footman, ties a rope about his bare neck and draws him up by pulleys to the ceiling, and there keeps him hanging as long as his master or the company pleases – the said footman, to the wonder and delight of all beholders, with a pot of ale in one hand and a pipe in the other – and when he is let down, there will not appear the least mark of the cord about his neck.

He bids a lady's maid put her finger into a cup of clear liquor, like water, upon which her face and both her hands are immediately withered like an old woman of fourscore, her belly swells as if she were within a week of her time, and her legs are as thick as mill posts; but upon putting her finger into another cup, she becomes as young and handsome as she was before.

He gives any gentleman leave to drive forty twelvepenny nails up to the head in a porter's backside, and then places the said porter on a lodestone chair, which draws out every nail, and the porter feels no pain.

He likewise draws the teeth of half a dozen gentlemen, mixes and jumbles them in a hat, gives any person leave to blindfold him and returns each their own, and fixes them as well as ever.

With his forefinger and thumb he thrusts several gentlemen's and ladies's eyes out of their heads without the least pain – at which time they see an unspeakable number of beautiful colours; and after they are entertained to the full, he places them again in their proper sockets without any damage to the sight.

He lets any gentleman drink a quart of hot melted lead, and by a draught of prepared liquor, of which he takes part himself, he makes the said lead pass through the said gentleman before all the spectators without any damage. After which it is produced in a cake to the company.

With many other wonderful performances of art too tedious here to mention.

The said artist has performed before most kings and princes in Europe with great applause.

He performs every day (except Sundays) from ten of the clock in the forenoon to one, and from four till seven in the evening, at the New Inn in Smithfield.

The first seat a British crown, the second a British half-crown, and the lowest a British shilling.

NB: The best hands in town are to play at the said show.

Note on the Texts and Notes

The text of *The Grand Mystery* (1726) is taken from the second edition, published in London in 1726. *The Wonderful Wonder of Wonders* (1720) is based on the sixth edition of 1722, which includes a preface and notes. *The Blunderful Blunder of Blunders* has been reprinted from the first edition, published in London in 1721. *Serious and Cleanly Meditations* and *The Bog House, A Poem* are reproduced from the 1723 London edition. The text of *The Bog House and Glass-Window Miscellany* is from the 1744 edition of Swift's *The Benefit of Farting Explained* (which also includes *The Wonderful Wonder of Wonders*, *Serious and Cleanly Meditations* and *The Bog House, A Poem*). *The Wonder of All the Wonders* is based on the 1722 edition, published in London.

The Grand Mystery

p. 3, *Dr W—d*: The physician and natural historian John Woodward (1665–1728), who developed a new theory of the earth which tried to explain fossils and other geological findings in the light of biblical history. He was the subject of numerous satires and lampoons, and a favourite target for John Arbuthnot and the Scriblerus Club.

p. 3, *Pyrmont*: A German mineral water which was very fashionable in England in the early eighteenth century.

p. 7, *B—tl–y*: The philologist and classical scholar Richard Bentley (1662–1742). He became master of Trinity College, Cambridge, where he published or commissioned many important scholarly works, including his own edition of Horace's poetry (1711).

p. 17, *Brutus's temper... Casca's dagger in a t—d*: Brutus, Cassius and Casca were three of the conspirators who took part in Caesar's assassination in 44 BC.

p. 17, *Edmund Ironsides*: Edmund II (*d*.1016), King of the English. The epithet "Ironsides" refers to his efforts in warding off a Danish invasion led by King Canute.

p. 18, *was basely killed… tribute to nature*: The chronicler Geoffrey Gaimar (*fl*.1136) tells the story of Edmund Ironsides being murdered in the privy by the sons of Edric Streona (*d*.1017).

p. 21, *Constitutions of the Freemasons*: *The Constitutions of the Freemasons. Containing the history, charges, regulations, etc. of that fraternity*, a 1721 work compiled by the historical writer and Church of Scotland Minister James Anderson (1679–1739).

p. 22, *oracles in Moorfields… D–nc–n C–mpb–ll*: Moorfields in London was renowned for being the home to astrologers and fortune tellers, and Duncan Campbell (*c*.1680–1730) was a notorious soothsayer and vendor of miraculous medicines.

p. 24, *protocaccographer*: A pioneer in the study of faeces.

p. 30, *Mahometans*: Turks.

p. 33, *whited-brown paper*: A coarse brown paper.

p. 34, *B—t's legend*: Possibly a reference to Gilbert Burnet, (1643–1715), bishop of Salisbury and historian.

p. 34, *Den—'s or A—e Ph—s's late easy labours*: A reference to the literary critic and translator John Dennis (1658–1734) and the poet and playwright Ambrose Philips (1674–1749).

p. 34, *Dr Marten's or Dr Cam's treatises*: The surgeons John Marten and Joseph Cam, who were active in the first two decades of the eighteenth-century and wrote a number of treatises on venereal diseases.

p. 34, *Society of Cripplegate Grey Beards… Father Jacob's pulpit declamations*: A reference to Joseph Jacob (*c*.1667–

1722), an independent minister who established a church distinguished by a strict code of conduct and some exclusive rites and rules such as the compulsory growing of moustaches for men. His sermons were fiercely political, and he was more than once abandoned by his congregation. Towards 1710 he moved to Curriers' Hall, Cripplegate, where he founded a new dissenting church.

p. 36, *draught*: "Drawing", but a pun may be intended, as the word "draught" was also used for a cesspool or a privy.

p. 37, *saltpetre… geneva*: Saltpetre is another term for potassium nitrate, a compound used in the manufacture of gunpowder and made from urine. Geneva, a cordial distilled from grain, was used as a stimulant and diuretic.

The Wonderful Wonder of Wonders

p. 41, *Ars enim non habet inimicum nisi ignorantem*: "Art has no enemy but ignorance" (Latin).

p. 43, *one of his heroes… squeak*: It is not clear who the hero "caught turning his own spit" is, although it may refer to Achilles dealing with the meat in *Iliad* xxiv, 626. The hero "surprised in his nudities" is Ulysses, watched by Nausicaa as he wakes up by the riverside (*Odyssey* vi).

p. 48, *gold-finders*: Originally the term "gold-finder" indicated a scavenger who looked "into jakes for bits o' money"; subsequently it came to be used for anybody who removed waste and excrement from people's houses and cesspools.

p. 48, *Magister artis, ingeniique largitor venter*: "Necessity is the fertile mother of all invention" (Latin).

p. 49, *the story of Bel and the Dragon*: A reference to the biblical tale included in the extended version of the Book of Daniel.

p. 50, *helluo librorum*: A "devourer of books", a "bookworm" (Latin).

p. 50, *Jacobus de Voragine*: The famous chronicler and Archbishop of Genoa Jacopo da Varazze (*c*.1230–98), author of *The Golden Legend*, a collection of saints' lives. Here the Latin name is used to pun on the other meaning of Voragine: "chasm" or "pit".

p. 51, *Rump Parliament*: Here the author is punning on the other, more common meaning of the word "rump".

p. 52, *Bembo's Lives of the Cardinals*: A reference to *The History of Venice* by the Italian scholar, poet and cardinal Pietro Bembo (1470–1547).

The Blunderful Blunder of Blunders

p. 63, *Jo. D——r*: Possibly the newsletter writer John Dyer (*c*.1653–1713).

p. 64, *a b——p*: A bishop.

p. 64, *the Wooden Man in Essex Street*: An allusion to a figure ornamenting a tobacconist's shop in Essex Street, Dublin.

p. 65, *Pretender*: James Francis Edward Stuart (1688–1766), son of the Catholic James II (1633–1701), the last of the Stuart monarchs, who was deposed in the "Glorious Revolution" of 1688. Those in support of a Stuart restoration became known as "Jacobites". When James II died in 1701, his son styled himself James III of England and James VIII of Scotland, and became known as "the Old Pretender". He was succeeded in his turn by his son Charles Edward, "the Young Pretender", whose defeat at the Battle of Culloden in 1745 brought to an end the Jacobite cause.

p. 65, *non quia nasus cœteris nullus erat*: A slightly adapted quote from Horace's *Satires* II, 2, 89–90, "Not because the others had no noses".

p. 66, *the losses sustained by the South Sea*: A reference to the South Sea Bubble of 1720, in which speculation in the South Sea Company's stock led to financial collapse.

Serious and Cleanly Meditations

p. 67, *Gold-finders*: See first note to p. 48.

p. 76, *cover-sluts*: An outer garment; an apron or pinafore.

p. 77, *C—l*: The bookseller Edmund Curll (*c*.1675–1747), who was often satirized by contemporary authors, including Pope and Swift, for publishing their works without permission.

p. 79, *St Austin*: St Augustine.

p. 79, *Duns Scoti… Weekly Journal*: A mixture of famous, obscure and ephemeral publications, from the *Complete Works* of the theologian Duns Scotus (*c*.1266–1308) to the scandalistic *Half-penny Post* and *Weekly Journal*.

p. 81, *Whom not… th' oppressed*: From Pope's *Windsor Forest*, 315–18. Belerium was the Roman name for the Penwith peninsula in Cornwall.

p. 83, *Phileleutherus Lipsiensis*: A pseudonym used by the scholar Richard Bentley (see note to p. 7).

p. 84, *Sir James Baker*: Probably a fictitious character. The same pseudonym was used by Pope in a poem included in his *A Key to the Lock* (1715) and in a 1717 pamphlet sometimes attributed to him.

p. 84, *Mr Tho. D—f–y, or Mr A—n H–ll*: The poet and playwright Thomas D'Urfey (*c*.1653–1723) and the writer and entrepreneur Aaron Hill (1685–1750).

p. 86, *Omnibus hoc vitium est cantoribus atque poetis*: A slightly adapted quote from Horace's *Satires* I, 3, 1: "This is a fault common to all singers and poets".

The Bog House, A Poem

p. 93, *adust*: "Scorched, dried up with heat, parched".

p. 93, *Alcides… stable rinsed*: In Greek mythology, one of the twelve labours of Hercules (Alcides) was to clean, in a single day, the stables of King Augeas, which contained thousands of cattle and had never been cleaned before. In order to complete the task, Hercules diverted a river through the stables.

p. 94, *Moloch*: A god to whom child sacrifices were made by the ancient Hebrews.

p. 94, *immitted*: "Inserted, introduced", the opposite of "emitted".

p. 95, *abstersive*: A purgative.

p. 96, *Jerne*: One of the ancient names of Ireland.

p. 96, *S——*: Possibly Satan.

p. 96, *medicinal wood*: Rhubarb.

The Bog House and Glass-window Miscellany

p. 99, *Dr Ch——ne of Bath*: The physician George Cheyne (*c.*1671–1743).

p. 99, *B——b D——n*: The politician and diarist George Bubb Dodington, Baron Melcombe (*c.*1690–1762).

p. 100, *Mr A——n H—ll, or the Laureate*: See note to p. 84. "The Laureate" probably refers to Colley Cibber (1671–1757), who was Poet Laureate from 1730 until his death.

p. 101, *B——ns*: Probably "Burns", although the reference cannot be traced with certainty.

p. 101, *T.C. the player*: Probably the actor and theatre manager Thomas Chapman (*c.*1683–1747).

p. 104, *c——y*: "Cony", slang for vagina.

p. 107, *Quæ… miracula?*: "What greater wonders than these do you think there are?" (Latin).

Biographical Note

JONATHAN SWIFT was born in Dublin on 30th November 1667, the second and only male child in a family of Anglo-Irish protestants. His father had died some seven months prior to his birth and, following his mother's return to England, Swift was left from a young age in the care of relatives. In 1673, at the age of six, he began his education at the prestigious Kilkenny Grammar School. In 1682 he entered Trinity College, Dublin, graduating in 1686. The violence surrounding William of Orange's accession to the throne in the "Glorious Revolution" of 1688 brought a temporary halt to Swift's academic career, forcing him to abandon studies for his Master's degree. He found refuge in England, where his mother's connections enabled him to secure a position as an assistant to Sir William Temple. During his residence at Moor Park, Temple's Surrey residence, Swift encountered the then eight-year-old Esther Johnson, later to become the focus of his "Stella" poems and the first in a number of intense but seemingly platonic female friendships that Swift experienced over the course of his life. It was also during this period that Swift first began to display symptoms of Ménière's disease. On the advice of his doctors Swift returned to Ireland in 1690 but returned to Moor Park shortly afterwards. In 1691 he visited Oxford and in 1692 received his MA from Hertford College. In the same year his first poem was published, although Swift's ambitions seem to have remained directed at furthering his career in the Anglican church. With this in mind he returned to Ireland in 1694 to take on the post of prebendary in the isolated parish of Kilroot, County Antrim. He returned to Moor Park in 1696, where over the next three years he composed both *A Tale of a Tub* and *The Battle of the Books*. Temple's death in 1699 saw Swift return to Ireland again, taking up a clerical position at Laracor. Over the next few years Swift's literary output and his reputation blossomed. He received his Doctorate in Divinity from Trinity College in 1702, and in 1704 both *A Tale of a Tub* and *The Battle of the Books* were published. This period saw the beginning of his lifelong friendships with fellow writers Pope, Gay and Arbuthnot and the consequent formation of the Scriblerus Club in 1713. He also began to emerge as a figure in the Whig politics of the day, particularly as an advocate for Ireland. This initiated the production of a number of satirical pamphlets, including the *Drapier's Letters* (1724) and *A Modest Proposal* (1729), and his most famous satirical work *Gulliver's Travels* (1726). Swift's final visit to England took place in 1727, and following the death of Esther Johnson in 1728 his work became increasing morbid. His self-written obituary in the form of *Verses on the Death of Dr Swift* was published in 1739 and his health began to decline. He died on 19th October 1745, and was buried in St Patrick's Cathedral Dublin next to Esther Johnson.